Tucholsky   Wagner   Zola   Scott   Schlegel
     Turgenev   Wallace   Fonatne Sydow   Freud
         Twain   Walther von der Vogelweide Fouqué   Friedrich II. von Preußen
                         Weber   Freiligrath   Frey
                                      Kant   Ernst
  Fechner   Fichte Weiße Rose von Fallersleben   Richthofen   Frommel
                                       Hölderlin
              Engels   Fielding   Eichendorff   Tacitus   Dumas
     Fehrs   Faber   Flaubert
                                              Eliasberg   Ebner Eschenbach
     Feuerbach   Maximilian I. von Habsburg Fock   Eliot   Zweig
                             Ewald   Vergil
              Goethe   Elisabeth von Österreich   London
   Mendelssohn Balzac   Shakespeare   Dostojewski   Ganghofer
                     Lichtenberg   Rathenau   Doyle   Gjellerup
           Trackl   Stevenson   Hambruch
    Mommsen   Tolstoi   Lenz   Droste-Hülshoff
              Thoma   Hanrieder
     Dach   Verne   von Arnim   Hägele   Hauff   Humboldt
              Reuter   Rousseau   Hagen   Hauptmann   Gautier
         Karrillon   Garschin   Baudelaire
              Damaschke   Defoe   Hebbel
                                  Descartes   Hegel   Kussmaul   Herder
    Wolfram von Eschenbach   Dickens   Schopenhauer   Rilke   George
         Bronner   Darwin   Melville   Grimm Jerome   Bebel
            Campe   Horváth   Aristoteles   Proust
      Bismarck   Vigny   Barlach   Voltaire Federer   Herodot
                         Gengenbach   Heine
         Storm   Casanova   Tersteegen   Grillparzer   Georgy
                       Chamberlain Lessing   Gilm
         Brentano   Langbein   Gryphius
                    Claudius   Schiller   Lafontaine
          Strachwitz   Kralik   Iffland Sokrates
                                          Bellamy   Schilling
                Katharina II. von Rußland   Gerstäcker   Raabe   Gibbon   Tschechow
       Löns   Vulpius
             Hesse   Hoffmann   Gogol   Wilde   Gleim
            Luther   Heym   Hofmannsthal   Klee   Hölty   Morgenstern   Goedicke
                Roth   Heyse   Klopstock   Kleist
    Luxemburg   Puschkin   Homer   Mörike   Musil
                 La Roche   Horaz
         Machiavelli   Kierkegaard   Kraft   Kraus
   Navarra   Aurel   Musset   Moltke
                                  Lamprecht   Kind
           Nestroy   Marie de France   Kirchhoff   Hugo
                                       Laotse   Ipsen   Liebknecht
           Nietzsche   Nansen
                     Marx   Lassalle   Gorki   Klett   Ringelnatz
               von Ossietzky   Leibniz
                          May   vom Stein   Lawrence   Irving
             Petalozzi
                  Platon   Knigge
                        Pückler   Michelangelo   Kafka
               Sachs   Poe   Kock
                                                    Liebermann
                         de Sade Praetorius   Mistral   Korolenko
                                                Zetkin

The publishing house tredition has created the series **TREDITION CLASSICS**. It contains classical literature works from over two thousand years. Most of these titles have been out of print and off the bookstore shelves for decades.

The book series is intended to preserve the cultural legacy and to promote the timeless works of classical literature. As a reader of a **TREDITION CLASSICS** book, the reader supports the mission to save many of the amazing works of world literature from oblivion.

The symbol of **TREDITION CLASSICS** is Johannes Gutenberg (1400 – 1468), the inventor of movable type printing.

With the series, tredition intends to make thousands of international literature classics available in printed format again – worldwide.

All books are available at book retailers worldwide in paperback and in hardcover. For more information please visit: www.tredition.com

tredition was established in 2006 by Sandra Latusseck and Soenke Schulz. Based in Hamburg, Germany, tredition offers publishing solutions to authors and publishing houses, combined with worldwide distribution of printed and digital book content. tredition is uniquely positioned to enable authors and publishing houses to create books on their own terms and without conventional manufacturing risks.

For more information please visit: www.tredition.com

Opening Ceremonies of the New York and Brooklyn Bridge, May 24, 1883

William C. Kingsley

Imprint

This book is part of the TREDITION CLASSICS series.

Author: William C. Kingsley
Cover design: toepferschumann, Berlin (Germany)

Publisher: tredition GmbH, Hamburg (Germany)
ISBN: 978-3-8491-6570-3

www.tredition.com
www.tredition.de

Copyright:
The content of this book is sourced from the public domain.

The intention of the TREDITION CLASSICS series is to make world literature in the public domain available in printed format. Literary enthusiasts and organizations worldwide have scanned and digitally edited the original texts. tredition has subsequently formatted and redesigned the content into a modern reading layout. Therefore, we cannot guarantee the exact reproduction of the original format of a particular historic edition. Please also note that no modifications have been made to the spelling, therefore it may differ from the orthography used today.

INTRODUCTORY.

The New York and Brooklyn Bridge was formally opened on Thursday, May 24th, 1883, with befitting pomp and ceremonial, in the presence of the largest multitude that ever gathered in the two cities. From the announcement by the Trustees of the date which was to mark the turning-over of the work to the public, it was evident that the popular demonstration would be upon a scale commensurate with the magnificence of the structure and its importance to the people of the United States. The evidences of widespread and profound interest in the event were early and unmistakable. They were not confined to the metropolis and its sister city on the Long Island shore, nor yet to the majestic Empire State. The occurrence was recognized as one of National importance; and throughout 8 the Union, from the rocky headlands of Maine to the golden shores of the Pacific, and from the gleaming waters of the St. Lawrence to the vast expanse of the Mexican Gulf, the opening ceremonies were regarded with intelligent concern and approval. Nearly every State contributed its representatives to the swelling throng that attended, while those who were unable to be present contemplated with pride and satisfaction the completion and consecration to its purpose of the greatest engineering work of modern times.

In the communities most directly benefited by the Bridge the demonstration was confined to no class or body of the populace. It was a holiday for high and low, rich and poor; it was, in fact, the People's Day. More delightful weather never dawned upon a festal morning. The heavens were radiant with the celestial blue of approaching summer; silvery fragments of cloud sailed gracefully across the firmament like winged messengers, bearing greetings of work well done; the clearest of spring sunshine tinged everything with a touch of gold, and a brisk, bracing breeze blown up from the Atlantic cooled the atmosphere 9 to a healthful and invigorating temperature. The incoming dawn revealed the twin cities gorgeous in gala attire. From towering steeple and lofty façade, from the fronts of business houses and the cornices and walls of private dwellings, from the forests of shipping along the wharves and the vessels in the dimpled bay, floated bunting fashioned in every conceivable design, while high above all, from the massive and enduring granite towers of the Bridge the Stars and Stripes signaled to the world from the gateway of the continent the arrival of the auspicious day.

Almost before the sun was up the thoroughfares of both cities put on a festival appearance. Business was generally suspended. The mercantile and professional communities vied with one another in the extent and splendor of their decorations, while from the hearty voice of Labor arose a chorus of ringing acclamation. Tens of thousands of men, women and children crowded into the streets, and, after gazing admiringly upon the decorations, wended their way in the direction of the mighty river span. From neighboring cities and from the adjacent country for many miles 10 around the incoming trains brought multitudes of excursionists and sight-seers. It seemed marvelous that they could all find accommodation, but the generous hospitality of the cities was cordially extended, and all were adequately provided for. The scenes presented during the day upon the streets and avenues of New York and Brooklyn will never be forgotten by those who witnessed them. Notwithstanding the enormous massing of people, the best of order was everywhere observable, and the day happily was free from any accident of a serious nature. The arrangements for the celebration were of a sensible and becoming character, and beside insuring an unobstructed and speedy course for the ceremonies, contributed beyond measure to the popular enjoyment.

Early in the afternoon the President of the United States, Gen. Chester A. Arthur, and the Hon. Grover Cleveland, Governor of the State of New York, the former accompanied by the members of his Cabinet and the latter by the officers of his Staff, were escorted from the Fifth Avenue Hotel to the New York City 11 Hall, where they were joined by his Honor Mayor Franklin Edson and the New York officials. From the City Hall the procession proceeded to the New York Approach to the Bridge. The Seventh Regiment, N.G., S.N.Y., Col. Emmons Clark, commanding, acted as escort to the Presidential and Gubernatorial party. The regimental band, of 75 pieces, headed the column and played popular airs as the procession moved along the crowded and gaily decorated thoroughfares. At the New York Tower a battalion of the Fifth United States Artillery, under command of Major Jackson, joined the escort, and between the lines of brilliantly uniformed troops the distinguished guests passed upon the roadway. They were formally received by a Committee of the Bridge Trustees, headed by Mr. William C. Kingsley, Vice-President and acting President of the Board.

The arrival at the New York Tower was proclaimed to the multitudes on shore by the thundering of many cannon. Salutes were fired from the forts in the harbor, from the United States Navy Yard, and from the summit of Fort 12 Greene. The United States fleet, consisting of the "Tennessee," the "Yantic," the "Kearsarge," the "Vandalia," and the "Minnesota," Rear-Admiral George H. Cooper, commanding, was anchored in the river below the Bridge and joined in the salute. As the procession moved across the roadway the yards of the men-of-war were manned, and from the docks and factories arose a tremendous babel of sounds, caused by the clanging of bells, the roaring of steam whistles, and the cheers of enthusiastic people, while sounding from afar, in delightful contrast with the clamorous discord, the silver chimes of Trinity rang out upon the river.

In the ornate iron railway depot at the Brooklyn terminus, where the exercises were to take place, the arrival of the approaching procession was anxiously awaited. The interior was bright with tasteful decorations, the prevailing feature being the sky-blue hangings of satin bordered with silver, and the coats-of-arms of the States appropriately interspersed amid a forest of flags. On the Brooklyn side the duties of escort were transferred to the 23d Regiment, N.G., S.N.Y., 13 Colonel Rodney C. Ward commanding. The regiment appeared upon this occasion for the first time in their new State service uniform, and performed their duties most efficiently. The arrangements for the procession and exercises were under the direction of Major-General James Jourdan, commanding the Second Division, N.G., S.N.Y., who was ably assisted by the members of the Division Staff. The building was thronged in every part. In the throng were many of the most conspicuous citizens of New York and other States, including representatives of the bench, the bar, the pulpit, the press, and all other professions. Beside the President and his Cabinet, consisting of the Hon. Charles J. Folger, Secretary of the Treasury; the Hon. William E. Chandler, Secretary of the Navy; the Hon. Henry M. Teller, Secretary of the Interior; the Hon. Walter Q.

Gresham, Postmaster-General, and the Hon. Benjamin Harris Brewster, Attorney-General; and Governor Cleveland and Staff, there were present the Governors of several States and the Mayors of nearly all the cities in the vicinity of the 14 metropolis. In the vast assemblage none were more conspicuous than the officers of the Army and Navy, who occupied an entire section and attracted general attention.

When the Presidential party and their escort entered the hall they were greeted with enthusiastic cheers. They occupied seats directly opposite the stand erected for the orators of the day. The exercises proceeded without delay in an orderly manner, and were appropriate and impressive throughout. Music was furnished during the ceremonies by the bands of the Seventh and Twenty-third regiments. The Hon. James S.T. Stranahan presided with the skill and dignity gained during his long experience in public life. Near him were the speakers, Mr. William C. Kingsley, Rev. Richard S. Storrs, D.D., the Hon. Abram S. Hewitt, Mayor Franklin Edson, of New York, and Mayor Seth Low, of Brooklyn, together with the members of the Board of Bridge Trustees. Mr. Stranahan opened the ceremonies by introducing Bishop Littlejohn, who wore the Episcopal robes. The Bishop fervently and impressively made the opening prayer, the great 15 assemblage bowing their heads reverentially during its delivery. Vice-President Kingsley was next introduced, and was received with hearty applause. Mr. Kingsley, in clear and distinct tones, and in comprehensive and business-like terms, proceeded to make the formal speech presenting the Bridge to the cities of New York and Brooklyn. The address was heard with careful attention, and upon its conclusion a round of enthusiastic applause swept through the building. His Honor Mayor Low followed Mr. Kingsley with a concise and appropriate speech, receiving the structure on behalf of the City of Brooklyn. His address elicited several demonstrations of approval from the audience. The Hon. Franklin Edson, Mayor of New York, who was the next speaker, was heartily applauded as he aptly accepted the Bridge in behalf of the authorities of the great metropolis. When Mr. Hewitt was introduced as the orator on the part of New York City, he was warmly cheered. His eloquent address riveted the attention of his hearers from beginning to end, and his pointed and conclusive vindication of the bridge manage 16 ment from the outset aroused the enthusiasm of his hearers to the utmost pitch. Following Mr. Hewitt came the Rev. Richard S. Storrs, D.D., who delivered the oration on behalf of Brooklyn. Never did the distinguished preacher appear to better advantage, and his oration, which was punctuated with applause, was characterized as a masterpiece by all who heard it. Upon the conclusion of his address the presiding officer declared the exercises at an end, and the company in the building dispersed.

The festivities, however, did not end with the conclusion of the formal ceremonies. The celebration was continued in both cities throughout the day and far into the night. Thousands upon thousands of enthusiastic people crowded the streets. After the ceremonies, the President, the Governor, the speakers of the day, and the Trustees were driven to the residence of Col. Washington A. Roebling, on Columbia Heights, where a reception was held. As they passed through the streets the people cheered as people only can who cheer in the atmosphere of a free government. From Col. 17 Roebling's house the company proceeded to the residence of Mayor Low, where they were entertained at a banquet. In the evening, under the auspices of the Municipal authorities, a grand reception to President Arthur and Governor Cleveland was given by the citizens of Brooklyn at the Academy of Music, and was attended by a great multitude. Another striking feature of the celebration at night was the display of fireworks on the Bridge given under the direction of the Board of Trustees. The pyrotechnic exhibition was viewed by almost the entire populace of the two cities, and a vast concourse of visitors from abroad. The East River was fairly blocked with craft of every description bearing legions of delighted spectators, and the streets and housetops were packed with people. The display was generally characterized as one of the grandest ever witnessed in America. The people of both cities evinced their public spirit in the decorations by day and the illuminations by night. The illuminations in Brooklyn, particularly, were on a magnificent scale, and excited the admiration of multitudes 18 of visitors to the city. In addition to the special features of the celebration there were many entertainments in honor of the event, including concerts in the various city parks. Throughout the afternoon and evening the best of order was preserved; the casualties that occurred were few and unimportant, and the auspicious day ended without the intrusion of anything that would carry with it other than pleasant memories of the significant event which it commemorated.

19

Order of Religious Services,

Conducted by Rt. Rev. A.N. Littlejohn, D.D.

The eternal God is thy refuge, and underneath are the everlasting arms. Deut. xxxiii.: 27.

Know therefore that the Lord thy God, He is God, the faithful God, which keepeth covenant and mercy with them that love Him and keep His commandments to a thousand generations. Deut. vii.: 9.

Remember the marvelous works that He hath done: His wonders, and the judgments of his mouth. Psalm cv.: 5.

Marvelous things did He in the sight of our forefathers, in the land of Egypt, even in the field of Zoan.

He divided the sea, and let them go through: He made the waters to stand on an heap.

In the day time also He led them with a cloud, and all the night through with a light of fire. Psalm lxxviii.: 13, 14, 15. 20

Oh, that men would therefore praise the Lord for His goodness and declare the wonders that He doeth for the children of men. Psalm cvii.: 21.

The Lord hath been mindful of us, and He shall bless us; He shall bless them that fear the Lord, both small and great. Psalm cxv.: 12, 13.

Glory be to the Father, and to the Son, and to the Holy Ghost:

As it was in the beginning, is now, and ever shall be, world without end.

Praise ye the Lord:

The Lord's name be praised.

PRAYER.

Almighty God, who hast in all ages showed forth Thy power and mercy in the preservation and advancement of the race redeemed by the precious blood of Thy dear Son: we yield Thee our unfeigned thanks and praise as for all Thy public mercies, so especially for the signal manifestation of Thy Providence which we commemorate this day. All things—wealth, industry, energy, skill, genius—come of Thee; and when we consecrate their triumphs unto Thee, 21 we give Thee but Thine own. Enable us to see in the strength and grandeur of this structure the evident tokens of Thy power, bringing mighty things to pass through the weakness of Thy creatures. Give us grace and wisdom to discern in all this work the nobler uses it was ordained by Thee to subserve. Teach us to know that all this mighty fabric is but vanity, save as it shall promote Thy sovereign purpose toward the sons of men. O Lord God, clothed with majesty and honor, decking Thyself with light as with a garment, and spreading out the heavens like a curtain, with the beams of Thy chambers in the waters, and the clouds for Thy chariot, walking upon the wings of the wind, Thy messengers spirits and Thy ministers a flaming fire, accept, we beseech Thee, this last and chiefest fruit of human toil and genius as a tribute to Thy glory, and a new power making for righteousness and peace amid all conflicts of earthly interests, and all the stir and pomp of worldly aggrandizement. Our life is a thing of nought, and our purposes vanish away; but 22 Thy years shall not fail, and with Thee the beginning and the end are the same. Therefore we implore Thee to bless and direct this work, that it shall be more than a highway for the things that perish, even a path of Thy eternal Spirit lifting by His own infinite grace, more and more, as the years roll on, the people of these cities toward the plane of Thine own life—the life of endless peace, of absolute unity, and perfect love, through Jesus Christ, the one Redeemer and Mediator between God and man. Amen.

23

Address of Wm. C. Kingsley,

President of the Board of Trustees.

In the presence of this great assemblage, and of the chosen representatives of the people of these two great cities, of the Governor of the State of New York and of the President of the United States, the pleasing duty devolves upon me, as the official agent of the Board of Trustees of the New York and Brooklyn Bridge, to announce formally to the chief magistrates of these two municipalities that this Bridge is now ready to be opened for public use, and is subject in its control and management only to such restrictions as the people, to whom it belongs, may choose to impose upon themselves. If I were at liberty to consult my own wishes I should not attempt to occupy your attention any further. I am not here as the spokesman of 24 my associates in the Board of Bridge Trustees. They are well content to let this great structure speak for them, and to speak more fittingly and more eloquently yet for the skillful, faithful and daring men who have given so many years of their lives—and in several instances even their lives—to the end that the natural barrier to the union, growth and greatness of this great commercial centre should be removed, and that a vast scientific conception should be matched in the skill, and courage, and endurance upon which it depended for its realization. With one name, in an especial sense, this Bridge will always be associated—that of Roebling. At the outset of this enterprise we were so fortunate as to be able to secure the services of the late John A. Roebling, who had built the chief suspension bridges in this country, and who had just then completed the largest suspension bridge ever constructed up to that time. His name and achievements were of invaluable service to this enterprise in its infancy. They secured for it a confidence not otherwise obtainable. He entered promptly and with more than professional zeal 25 into the work of erecting a bridge over the East River. As is universally known, while testing and perfecting his surveys his foot was crushed between the planks of one of our piers; lockjaw supervened, and the man who designed this Bridge lost his life in its service. The main designs were, however, completed by the elder Roebling before he met his sad and untimely death. He was succeeded at once by his son, Colonel Washington A. Roebling, who had for years before shared in his father's professional confidences and labors. Here the son did not succeed the father by inheritance merely. The elder Roebling, according to his own statements, would not have undertaken the conduct of this work at his age—and he was independent of mere professional gain—if it were not for the fact, as he frequently stated, that he had a son who was entirely capable of building this Bridge. Indeed, the elder Roebling advised that the son, who was destined to carry on and complete the work, should be placed in chief authority at the beginning. The turning point—as determining the feasibility of this enterprise—was reached 26 down in the earth, and under the bed of the East River. During the anxious days and nights while work was going on within the caissons, Colonel Roebling seemed to be always on hand, at the head of his men, to direct their efforts, and to guard against a mishap or a mistake which, at this stage of the work, might have proved to be disastrous. The foundations of the towers were successfully laid, and the problem of the feasibility of the Bridge was solved. Colonel Roebling contracted the mysterious disease in the caissons which had proved fatal to several of the workmen in our employ. For many long and weary years this man, who entered our service young and full of life, and hope, and daring, has been an invalid and confined to his home. He has never seen this structure as it now stands, save from a distance. But the disease, which has shattered his nervous system for the time, seemed not to have enfeebled his mind. It appeared even to quicken his intellect. His physical infirmities shut him out, so to speak, from the world, and left him dependent largely on the society of his family, but it gave him for a com 27 panion day and night this darling child of his genius—every step of whose progress he has directed and watched over with paternal solicitude. Colonel Roebling may never walk across this Bridge, as so many of his fellow-men have done to-day, but while this structure stands he will make all who use it his debtor. His infirmities are still such that he who would be the centre of interest on this occasion, and even in this greatly distinguished company, is conspicuous by his absence. This enterprise was only less fortunate in securing an executive head than in obtaining scientific direction. For sixteen years together the late Hon. Henry C. Murphy stood for this work wherever it challenged the enmity of an opponent or needed an advocate, a supporter and a friend. He devised the legislation under which it was commenced. He staked in its inception a large portion of his private fortune on its success. He upheld its feasibility and utility before committees, and legislatures, and law courts, and in every forum of public discussion. For years he looked forward to this day to fittingly close the activities of a long, useful 28 and, in many respects, an illustrious career. It was not permitted him to see it, but he saw very near the end, and he lived long enough to realize, what is now admitted, that he was to the end of his days engaged in a work from which the name of the city he loved so well will never be disassociated, for it is a work the history of which will for all time be embraced in the records of the achievements of American enterprise and of American genius. I am sure I speak for the Board of Trustees in returning their thanks to all the professional gentlemen who have been in our employ — and especially to Messrs. Martin, Paine, Farrington, McNulty and Probasco. For the most part these men have been engaged on the Bridge from its commencement to its completion. It has always seemed to the Trustees as if the highest and the humblest workmen engaged on this work were alike influenced by the spirit of enterprise in which the Bridge had its origin. Men whose daily compensation was not more than sufficient to provide them and their families with their daily bread were at all times ready to take their 29 lives in their hands in the performance of the imperative and perilous duties assigned them. In the direct prosecution of the work twenty men lost their lives. Peace hath its victories, and it has its victims and its martyrs, too. Of the seven consulting engineers to whom the matured plans of the elder Roebling were submitted — all men of the highest eminence in their profession — three have passed away, and four are living to witness, in the assured success of this structure, the one ratification of their judgment which cannot be questioned.

It remains for me to say, in conclusion, that the two cities rose at all times to the level of the spirit of our time and country. Their citizens staked millions on what seemed to many to be an experiment — a structure, it was often said, that at its best would not be of any actual use. How solid it is; how far removed it is from all sense of apprehension; how severely practical it is in all its relations, and how great a factor in the corporate lives of these cities it is destined to be, we all now realize. This Bridge has cost many millions of dollars, and it has taken many 30 years to build it. May I say on this occasion that the people whom you represent (turning to where the Mayors of the two cities stood together) would not part with the

Bridge to-day for even twice or thrice its cost? And may I remind those who, not unnaturally, perhaps, have been disappointed and irritated by delays in the past, that those who enter a race with Time for a competitor have an antagonist that makes no mistakes, is subject to no interference and liable to no accident.

31

Address of Hon. Seth Low,

Mayor of the City of Brooklyn.

Gentlemen of the Trustees—With profound satisfaction, on behalf of the City of Brooklyn, I accept the completed Bridge. Fourteen times the earth has made its great march through the heavens since the work began. The vicissitudes of fourteen years have tried the courage and the faith of engineers and of people. At last we all rejoice in the signal triumph. The beautiful and stately structure fulfills the fondest hope. It will be a source of pleasure to-day to every citizen that no other name is associated with the end than that which has directed the work from the beginning—the name of Roebling. With all my heart I give to him who bears it now the city's acknowledgment and thanks.

Fourteen years ago a city of 400,000 people 32 on this side of the river heard of a projected suspension bridge with incredulity. The span was so long, the height so great, and the enterprise likely to be so costly, that few thought of it as something begun in earnest. The irresistible demands of commerce enforced these hard conditions. But Science said, "It is possible," and Courage said, "It shall be!" To-day a city of 600,000 people welcomes with enthusiasm the wonderful creation of genius. Graceful, and yet majestic, it clings to the land like a thing that has taken root. Beautiful as a vision of fairy-land it salutes our sight. The impression it makes upon the visitor is one of astonishment, an astonishment that grows with every visit. No one who has been upon it can ever forget it. This great structure cannot be confined to the limits of local pride. The glory of it belongs to the race. Not one shall see it and not feel prouder to be a man.

And yet it is distinctly an American triumph. American genius designed it, American skill built it, and American workshops made it. About 1837 the Screw Dock across the river, then 33 known as the Hydrostatic Lifting Dock, was built. In order to construct it the Americans of that day were obliged to have the cylinders cast in England. What a stride from 1837 to 1883—from the Hydrostatic Dock to the New York and Brooklyn Bridge!

And so this Bridge is a wonder of science. But in no less degree it is a triumph of faith. I speak not now of the courage of those who projected it. Except for the faith which removes mountains yonder river could not have been spanned by this Bridge. It is true that the material which has gone into it has been paid for; the labor which has been spent upon it has received its hire. But the money which did these things was not the money of those who own the Bridge. The money was lent to them on the faith that these two great cities would redeem their bond. So have the Alps been tunneled in our day; while the ancient prophecy has been fulfilled that faith should remove mountains. We justify this faith in us as we pay for the Bridge by redeeming the bond.

In the course of the construction of the 34 Bridge a number of lives have been lost. Does it not sometimes seem as though every work of enduring value, in the material as in the moral world, must needs be purchased at the cost of human life? Let us recall with kindness at this hour the work of those who labored here faithfully unto the death, no less than of that great army of men who have wrought, year in and year out, to execute the great design. Let us give our meed of praise to-day to the humblest workman who has here done his duty well, no less than to the great engineer who told him what to do.

The importance of this Bridge in its far-reaching effects at once entices and baffles the imagination. At either end of the Bridge lies a great city—cities full of vigorous life. The activities and the energies of each flow over into the other. The electric current has conveyed unchecked between the two the interchanging thoughts, but the rapid river has ever bidden halt to the foot of man. It is as though the population of these cities had been brought down to the riverside, year after year, there to 35 be taught patience; and as though, in this Bridge, after these many years, patience had had her perfect work. The ardent merchant, the busy lawyer, the impatient traveler—all, without distinction and without exception—at the river have been told to wait. No one can compute the loss of time ensuing daily from delays at the ferries to the multitudes crossing the stream. And time is not only money—it is opportunity. Brooklyn becomes available, henceforth, as a place of residence to thousands, to whom the ability to reach their places of business without interruption from fog and ice is of paramount importance. To all Brooklyn's present citizens a distinct boon is given. The certainty of communication with New York afforded by the Bridge is the fundamental benefit it confers. Incident to this is the opportunity it gives for rapid communication.

As the water of the lakes found the salt sea when the Erie Canal was opened, so surely will quick communication seek and find this noble Bridge, and as the ships have carried hither and thither the products of the mighty 36 West, so shall diverging railroads transport the people swiftly to their homes in the hospitable city of Brooklyn. The Erie Canal is a waterway through the land connecting the great West with the older East. This Bridge is a landway over the water, connecting two cities bearing to each other relations in some respects similar. It is the function of such works to bless "both him that gives and him that takes." The development of the West has not belittled, but has enlarged New York, and Brooklyn will grow by reason of this Bridge, not at New York's expense, but to her permanent advantage. The Brooklyn of 1900 can hardly be guessed at from the city of to-day. The hand of Time is a mighty hand. To those who are privileged to live in sight of this noble structure every line of it should be eloquent with inspiration. Courage, enterprise, skill, faith, endurance—these are the qualities which have made the great Bridge, and these are the qualities which will make our city great and our people great. God grant they never may be lacking in our midst. Gentlemen of the Trus 37 tees, in accepting the Bridge at your hands, I thank you warmly in Brooklyn's name for your manifold and arduous labors.

Address of Hon. Franklin Edson,

Mayor of the City of New York.

Mr. President—On behalf of the City of New York, I accept the great work which you now tender as ready for the public use of the two cities which it so substantially and, at the same time, so gracefully joins together.

The City of New York joyfully unites with the City of Brooklyn in extending to you, sir, and to those who have been associated with you, sincere congratulations upon the successful completion of this grand highway, establishing, as it does, an enduring alliance between these two great cities. Through the wisdom, energy, zeal and patience of yourself and your co-laborers in this vast enterprise, we are enabled this day to recognize the fact that a common and unbroken current flows through the veins of these two cities, which must add in no small degree to the strength, healthful growth and prosperity of both, and we believe that what has thus been joined together shall never be put asunder.

When, more than fifteen years ago, you, Mr. President, foresaw the advantages that would surely accrue to these cities from the establishment of such a means of communication between them, few could be found to look upon such advantages as other than, at best, problematical. To-day, however, they are recognized, and so fully, that before this Bridge was completed the building of another not far distant had begun to be seriously considered.

It was forty years after the vast advantages of water communication between the Hudson and the great lakes had dawned upon the mind of Washington, in the course of a tour through the valley of the Mohawk, that such a work came to be appreciated by the people, and resulted in that grand artery of wealth to our State, the Erie Canal. So I believe it has ever been in the past with the initiation and construction of great public works, and with the introduction of agencies and methods which have been of the greatest benefit to mankind throughout the world, and so perhaps it will ever be. Yet, for the welfare of these two cities, let us venture the hope that the tide of improvement and of active preparation is setting in, for it behooves us more than most are aware to be forecasting our future necessities, and to recognize the fact that

There is a tide in the affairs of *cities*,
Which, taken at the flood, leads on to fortune.

It is not difficult for most of us to look back twenty-five years and see clearly the wonderful strides which have been made in population, commerce, manufacturing and financial interests, and in all the industries which help to make great and prosperous communities; nor is it difficult to trace the wonders that have been wrought through the agencies of steam and electricity within those years. But to look forward twenty-five years and attempt to discern the condition of things in this metropolis, if they shall continue to move forward on the same scale of progress, is an undertaking that few can grasp. No one dares accept the possibilities that are forced upon the mind in the course of its contemplation. Will these two cities ere then have been consolidated into one great municipality, numbering within its limits more than five millions of people? Will the right of self-government have been accorded to the great city, thus united, and will her people have learned how best to exercise that right? Will the progress of improvement and the preparation for commerce, manufactories and trade, and for the comforts of home for poor and rich, have kept pace with the demand in the great and growing city? Will the establishment of life-giving parks, embellished with appropriate fountains and statues and with the numberless graces of art, which at once gladden the eye and raise the standard of civilization, have kept abreast with its growth in wealth and numbers?

These are but few of the pertinent questions which must be answered by the zealous and honest acts of the generation of men already in active life. Here are the possibilities; all the elements and conditions are here; but the results must depend upon the wisdom and patriotism and energy of those who shall lead in public affairs. May they be clothed with a spirit of wisdom and knowledge akin to that which inspired those who conceived and executed the great work which we receive at your hands and dedicate to-day.

Address of Hon. Abram S. Hewitt.

Two hundred and seventy years ago the good ship "Tiger," commanded by Captain Adraien Block, was burned to the water's edge, as she lay at anchor, just off the southern end of Manhattan Island. Her crew, thus forced into winter quarters, were the first white men who built and occupied a house on the land where New York now stands; "then," to quote the graphic language of Mrs. Lamb, in her history of the City, "in primeval solitude, waiting till commerce should come and claim its own. Nature wore a hardy countenance, as wild and as untamed as the savage landholders. Manhattan's twenty-two thousand acres of rock, lake and rolling table land, rising at places to a height of one hundred and thirty-eight feet, 44 were covered with sombre forests, grassy knolls and dismal swamps. The trees were lofty; and old, decayed and withered limbs contrasted with the younger growth of branches; and wild flowers wasted their sweetness among the dead leaves and uncut herbage at their roots. The wanton grapevine swung carelessly from the topmost boughs of the oak and the sycamore; and blackberry and raspberry bushes, like a picket guard, presented a bold front in all possible avenues of approach. The entire surface of the island was bold and granitic, and in profile resembled the cartilaginous back of the sturgeon."

This primeval scene was the product of natural forces working through uncounted periods of time; the continent slowly rising and falling in the sea like the heaving breast of a world asleep; glaciers carving patiently through ages the deep estuaries; seasons innumerable clothing the hills with alternate bloom and decay.

The same sun shines to-day upon the same earth; yet how transformed! Could there be a more astounding exhibition of the power of 45 man to change the face of nature than the panoramic view which presents itself to the spectator standing upon the crowning arch of the Bridge, whose completion we are here to-day to celebrate in the honored presence of the President of the United States, with their fifty millions; of the Governor of the State of New York, with its five millions; and of the Mayors of the two cities, aggregating over two millions of inhabitants? In the place of stillness and solitude, the footsteps of these millions of human beings; instead of the smooth waters "unvexed by any keel," highways of commerce ablaze with the flags of all the nations; and where once was the green monotony of forested hills, the piled and towering splendors of a vast metropolis, the countless homes of industry, the echoing marts of trade, the gorgeous palaces of luxury, the silent and steadfast spires of worship!

To crown all, the work of separation wrought so surely, yet so slowly, by the hand of Time, is now reversed in our own day, and "Manahatta" and "Seawanhaka" are joined again, as 46 once they were before the dawn of life in the far azoic ages.

"It is done!
Clang of bell and roar of gun
Send the tidings up and down.
How the belfries rock and reel!
How the great guns, peal on peal,
Fling the joy from town to town!"

"What hath God wrought!" were the words of wonder, which ushered into being the magnetic telegraph, the greatest marvel of the many marvelous inventions of the present century. It was the natural impulse of the pious maiden who chose this first message of reverence and awe, to look to the Divine Power as the author of a new gospel. For it was the invisible, and not the visible agency, which addressed itself to her perceptions. Neither the bare poles, nor the slender wire, nor the silent battery, could suggest an adequate explanation of the extinction of time and space which was manifest to her senses, and she could only say, "What hath God wrought!"

But when we turn from the unsightly telegraph to the graceful structure at whose portal 47 we stand, and when the airy outline of its curves of beauty, pendant between massive towers suggestive of art alone, is contrasted with the over-reaching vault of heaven above and the ever-moving flood of waters beneath, the work of omnipotent power, we are irresistibly moved to exclaim, "What hath *man* wrought!"

Man hath, indeed, wrought far more than strikes the eye in this daring undertaking, by the general judgment of engineers, without a rival among the wonders of human skill. It is not the work of any one man or of any one age. It is the result of the study, of the experience, and of the knowledge of many men in many ages. It is not merely a creation—it is a growth. It stands before us to-day as the sum and epitome of human knowledge; as the very heir of the ages; as the latest glory of centuries of patient observation, profound study and accumulated skill, gained, step by step, in the never-ending struggle of man to subdue the forces of nature to his control and use.

In no previous period of the world's history could this Bridge have been built. Within the 48 last hundred years the greater part of the knowledge necessary for its erection has been gained. Chemistry was not born until 1776, the year when political economy was ushered into the world by Adam Smith, and the Declaration of Independence was proclaimed by the Continental Congress, to be maintained at the point of the sword by George Washington. In the same year Watt produced his successful steam engine, and a century has not elapsed since the first specimen of his skill was erected on this continent. The law of gravitation was indeed known a hundred years ago, but the intricate laws of force, which now control the domain of industry, had not been developed by the study of physical science, and their practical applications have only been effectually accomplished within our own day, and, indeed, some of the most important of them during the building of the Bridge. For use in the caissons, the perfecting of the electric light came too late, though, happily, in season for the illumination of the finished work.

This construction has not only employed every abstract conclusion and formula of mathematics, 49 whether derived from the study of the earth or the heavens, but the whole structure may be said to rest upon a mathematical foundation. The great discoveries of chemistry, showing the composition of water, the nature of gases, the properties of metals; the laws and processes of physics, from the strains and pressures of mighty masses to the delicate vibrations of molecules, are all recorded here. Every department of human industry is represented, from the quarrying and the cutting of the stones, the mining and smelting of the ores, the conversion of iron into steel by the pneumatic process, to the final shaping of the masses of metal into useful forms, and its reduction into wire, so as to develop in the highest degree the tensile strength which fits it for the work of suspension. Every tool which the ingenuity of man has invented has somewhere, in some special detail, contributed its share in the accomplishment of the final result.

"Ah! what a wondrous thing it is
To note how many wheels of toil
One word, one thought can set in motion."

But without the most recent discoveries of science, which have enabled steel to be substituted for iron—applications made since the original plans of the Bridge were devised—we should have had a structure fit, indeed, for use, but of such moderate capacity that we could not have justified the claim which we are now able to make, that the cities of New York and Brooklyn have constructed, and to-day rejoice in the possession of, the crowning glory of an age memorable for great industrial achievements.

This is not the proper occasion for describing the details of this undertaking. This grateful task will be performed by the engineer in the final report, with which every great work is properly committed to the judgment of posterity. But there are some lessons to be drawn from the line of thought I have followed which may encourage and comfort us as to the destiny of man and the outcome of human progress.

What message, then, of hope and cheer does this achievement convey to those who would fain believe that love travels hand in hand with light along the rugged pathway of time? Have the discoveries of science, the triumphs of art and the progress of civilization, which have made its accomplishment a possibility and a reality, promoted the welfare of mankind, and raised the great mass of the people to a higher plane of life?

This question can best be answered by comparing the compensation of the labor employed in the building of this Bridge with the earnings of labor upon works of equal magnitude in ages gone by. The money expended for the work of construction proper on the

Bridge, exclusive of land damages and other outlays, such as interest, not entering into actual cost, is nine million ($9,000,000) dollars. This money has been distributed in numberless channels—for quarrying, for mining, for smelting, for fabricating the metals, for shaping the materials, and erecting the work, employing every kind and form of human labor. The wages paid at the Bridge itself may be taken as the fair standard of the wages paid for the work done elsewhere. These wages are: 52

	Average.	
Laborers,	$1 75	per day.
Blacksmiths,	3 50 to $4 00	do.
Carpenters,	3 00 to 3 50	do.
Masons and Stonecutters,	3 50 to 4 00	do.
Riggers,	2 00 to 2 50	do.
Painters,	2 00 to 3 50	do.

Taking all these kinds of labor into account, the wages paid for work on the Bridge will thus average $2.50 per day.

Now, if this work had been done at the time when the Pyramids were built, with the skill, appliances and tools then in use, and if the money available for its execution had been limited to nine million ($9,000,000) dollars, the laborers employed would have received an average of not more than two cents per day, in money of the same purchasing power as the coin of the present era. In other words, the effect of the discoveries of new methods, tools and laws of force, has been to raise the wages of labor more than an hundred fold, in the interval which has elapsed since the Pyramids were built. I shall not weaken the suggestive 53 force of this statement by any comments upon its astounding evidence of progress, beyond the obvious corollary that such a state of civilization as gave birth to the Pyramids would now be the signal for universal bloodshed, revolution and anarchy. I do not underestimate the hardships borne by the labor of our time. They are, indeed, grievous, and to lighten them is, as it should be, the chief concern of statesmanship. But this comparison proves that through forty centuries these hardships have been steadily diminished; that all the achievements of science, all the discoveries of art, all the inventions of genius, all the progress of civilization, tend by a higher and immutable law to the steady and certain amelioration of the condition of society. It shows that, notwithstanding the apparent growth of great fortunes, due to an era of unparalleled development, the distribution of the fruits of labor is approaching from age to age to more equitable conditions, and must, at last, reach the plane of absolute justice between man and man.

But this is not the only lesson to be drawn 54 from such a comparison. The Pyramids were built by the sacrifices of the living for the dead. They served no useful purpose, except to make odious to future generations the tyranny which degrades humanity to the level of the brute. In this age of the world such a waste of effort would not be tolerated. To-day the expenditures of communities are directed to useful purposes. Except upon works designed for defence in time of war, the wealth of society is now mainly expended in opening channels of communication for the free play of commerce, and the communion of the human race. An analysis of the distribution of the surplus earnings of man after providing food, shelter and raiment, shows that they are chiefly absorbed by railways, canals, ships, bridges and telegraphs. In ancient times these objects of expenditure were scarcely known. Our Bridge is one of the most conspicuous examples of this change in the social condition of the world, and of the feeling of men. In the Middle Ages cities walled each other out, and the fetters of prejudice and tyranny held the energies of man in hopeless bondage. To- 55 day men and nations seek free intercourse with each other, and much of the force of the intellect and energy of the world is expended in breaking down the barriers established by nature, or created by man, to the solidarity of the human race.

And yet, in view of this tendency, the most striking and characteristic feature of the nineteenth century, there still are those who believe and teach that obstruction is the creator of wealth; that the peoples can be made great and free by the erection of artificial barriers to the beneficent action of commerce, and the unrestricted intercourse of men and nations with each other. If they are right, then this Bridge is a colossal blunder, and the doctrine which bids us to love our neighbors as ourselves is founded upon a misconception of the divine purpose.

But the Bridge is more than an embodiment of the scientific knowledge of physical laws, or a symbol of social tendencies. It is equally a monument to the moral qualities of the human soul. It could never have been built by mere 56 knowledge and scientific skill alone. It required, in addition, the infinite patience and unwearied courage by which great results are achieved. It demanded the endurance of heat, and cold, and physical distress. Its constructors have had to face death in its most repulsive form. Death, indeed, was the fate of its great projector, and dread disease the heritage of the greater engineer who has brought it to completion. The faith of the saint and the courage of the hero have been combined in the conception, the design and the execution of this work.

Let us, then, record the names of the engineers and foremen who have thus made humanity itself their debtor for a successful achievement, not the result of accident or of chance, but the fruit of design, and of the consecration of all personal interest to the public weal. They are: John A. Roebling, who conceived the project and formulated the plan of the Bridge; Washington A. Roebling, who, inheriting his father's genius, and more than his father's knowledge and skill, has directed the execution of this great work from its inception to its completion; aided 57 in the several departments by Charles C. Martin, Francis Collingwood, William H. Paine, George W. McNulty, Wilhelm Hildenbrand and Samuel R. Probasco as assistant engineers; and as foremen by E.F. Farrington, Arthur V. Abbott, William Van der Bosch, Charles Young and Harry Tupple, who, in apparently subordinate positions, have shown themselves peculiarly fitted to command, because they have known how to serve. But the record would not be complete without reference to the unnamed men by whose unflinching courage, in the depths of the caissons, and upon the suspended wires, the work was carried on amid storms, and accidents, and dangers, sufficient to appall the stoutest heart. To them we can only render the tribute which history accords to those who fight as privates in the battles of freedom, with all the more devotion and patriotism because their names will never be known by the world whose benefactors they are. One name, however, which may find no place in the official records, cannot be passed over here in silence. In ancient times when great works were constructed, a goddess was chosen, to whose 58 tender care they were dedicated. Thus the ruins of the Acropolis to-day recall the name of Pallas Athene to an admiring world. In the Middle Ages, the blessing of some saint was invoked to protect from the rude attacks of the barbarians, and the destructive hand of time, the building erected by man's devotion to the worship of God. So, with this Bridge will ever be coupled the thought of one, through the subtle alembic of whose brain, and by whose facile fingers, communication was maintained between the directing power of its construction, and the obedient agencies of its execution. It is thus an everlasting monument to the self-sacrificing devotion of woman, and of her capacity for that higher education from which she has been too long debarred. The name of Mrs. Emily Warren Roebling will thus be inseparably associated with all that is admirable in human nature, and with all that is wonderful in the constructive world of art.

This tribute to the engineers, however, would not be deserved, if there is to be found any evidence of deception on their part in the origin 59 of the work, or any complicity with fraud in its execution and completion. It is this consideration which induced me to accept the unexpected invitation of the trustees to speak for the city of New York on the present occasion. When they thus honored me, they did not know that John A. Roebling addressed to me the letter in which he first suggested (and, so far as I am aware, he was the first engineer to suggest), the feasibility of a bridge between the two cities, so constructed as to preserve unimpaired the freedom of navigation. This letter, dated June 19, 1857, I caused to be printed in the *New York Journal of Commerce*, where it attracted great attention because it came from an engineer who had already demonstrated, by successfully building suspension bridges over the Schuylkill, the Ohio and the Niagara rivers, that he spoke with the voice of experience and authority. This letter was the first step towards the construction of the work, which, however, came about in a manner different from his expectations, and was finally completed on a plan more extensive than he had ventured to describe. It has been charged 60 that the original estimates of cost have been far exceeded by the actual outlay. If this were true, the words of praise which I have uttered for the engineers, who designed and executed this work, ought rather to have been a sentence of censure and condemnation. Hence, the invitation which came to me unsought, seemed rather to be an appeal from the grave for such vindication as it was within my power to make, and which could not come with equal force from any other quarter.

Engineers are of two kinds: the creative and the constructive. The power to conceive great works demands imagination and faith. The creative engineer, like the poet, is born, not made. If to the power to conceive, is added the ability to execute, then have we one of those rare geniuses who not only give a decided impulse to civilization, but add new glory to humanity. Such men were Michael Angelo, Leonardo da Vinci, Watt, Wedgwood, Brunel, Stephenson and Bessemer; and such a man was John A. Roebling. It was his striking peculiarity, that while his conceptions were bold and original, his execution 61 was always exact, and within the limits of cost which he assigned to the work of his brain. He had made bridges a study, and had declared in favor of the suspension principle for heavy traffic, when the greatest living authorities had condemned it as costly and unsafe. When he undertook to build a suspension bridge for railway use, he did so in the face of the deliberate judgment of the profession, that success would be impossible. Stephenson had condemned the suspension principle and approved the tubular girder for railway traffic. But it was the Nemesis of his fate, that when he came out to approve the location of the great tubular bridge at Montreal, he should pass over the Niagara River in a railway train, on a suspension bridge, which he had declared to be an impracticable undertaking.

When Roebling suggested the Bridge over the East River, his ideas were limited to the demands of the time, and controlled by the necessity for a profitable investment. He had no expectation that the two cities would embark in the enterprise. Indeed, in one of his letters 62 so late as April 14, 1860, he says, "As to the corporations of New York and Brooklyn undertaking the job, no such hope may be entertained in our time." In eight years thereafter, these cities had undertaken the task upon a scale of expense far exceeding his original ideas of a structure, to be built exclusively by private capital for the sake of profit.

How came this miracle to pass? The war of the rebellion occurred, delaying for a time the further consideration of Roebling's ideas. This war accustomed the nation to expenditures on a scale of which it had no previous conception. It did more than expend large sums of money. Officials became corrupt and organized themselves for plunder. In the city of New York, especially, the government fell into the hands of a band of thieves, who engaged in a series of great and beneficial public works, not for the good they might do, but for the opportunity which they would afford to rob the public treasury. They erected court-houses and armories; they opened roads, boulevards and parks; and they organized two of the grandest devices for trans portation which the genius of man has ever conceived; a rapid transit railway for New York, and a great highway between New York and Brooklyn. The Bridge was commenced, but the Ring was driven into exile by the force of public indignation, before the rapid transit scheme, since executed on a different route by private capital, was undertaken. The collapse of the Ring brought the work on the Bridge to a stand-still.

It was a timely event. The patriotic New Yorker might well have exclaimed, just before this great deliverance, in the words of the Consul of ancient Rome, in Macaulay's stirring poem,

"And if they once may win the bridge,
What hope to save the town?"

Meanwhile, the elder Roebling had died, leaving behind him his estimates and the general plans of the structure, to cost, independent of land damages and interest, about $7,000,000. This great work which, if not "conceived in sin," was "brought forth in iniquity," thus became the object of great suspicion, and of a prejudice which has not been removed to this day. I know that to many I make a startling announcement, when I state the incontrovertible fact, that no money was ever stolen by the Ring from the funds of the Bridge; that the whole money raised has been honestly expended; that the estimates for construction have not been materially exceeded; and that the excess of cost over the estimates is due to purchases of land which were never included in the estimates; to interest paid on the city subscriptions; to the cost of additional height and breadth of the Bridge; and the increase in strength rendered necessary by a better comprehension of the volume of traffic between the two cities. The items covered by the original estimate of $7,000,000 have thus been raised to $9,000,000, so that $2,000,000 represents the addition to the original estimates.

For this excess, amounting to less than thirty per cent., there is actual value in the Bridge in dimension and strength, whereby its working capacity has been greatly increased. The carriage-ways, as originally designed, would have permitted only a single line of vehicles in each direction. The speed of the entire procession, more than a mile long, would, therefore, have been limited by the rate of the slowest; and every accident causing stoppage to a single cart would have stopped everything behind it for an indefinite period. It is not too much to say that the removal of this objection, by widening the carriage-ways, has multiplied manifold the practical usefulness of the Bridge.

The statement I have made is due to the memory not only of John A. Roebling, but also of Henry C. Murphy, that great man who devoted his last years to this enterprise; and who, having, like Moses, led the people through the toilsome way, was permitted only to look, but not to enter upon the promised land.

This testimony is due also to the living trustees and to the engineers who have controlled and directed this large expenditure in the public service, the latter, in the conscientious discharge of professional duty; and the former, with no other object than the welfare of the public, and without any other possible reward than the good opinion of their fellow-citizens.

I do not make this statement without a full sense of the responsibility which it involves, and I realize that its accuracy will shortly be tested by the report of experts who are now examining the accounts. But it will be found that I have spoken the words of truth and soberness. When the Ring absconded I was asked by William C. Havemeyer, then the Mayor of New York, to become a trustee, in order to investigate the expenditures, and to report as to the propriety of going on with the work. This duty was performed without fear or favor. The methods by which the Ring proposed to benefit themselves were clear enough, but its members fled before they succeeded in reimbursing themselves for the preliminary expenses which they had defrayed. With their flight a new era commenced, and during the three years when I acted as a trustee, I am sure that no fraud was committed, and that none was possible. Since that time the Board has been controlled by trustees, some of whom are thorough experts in bridge building, and the others men of such high character that the suggestion of malpractice is improbable to absurdity. 67

The Bridge has not only been honestly built, but it may be safely asserted that it could not now be duplicated at the same cost. Much money might, however, have been saved if the work had not been delayed through want of means, and unnecessary obstacles interposed by mistaken public officials. Moreover, measured by its capacity, and the limitations imposed on its construction by its relation to the interests of traffic and navigation, it is the cheapest structure ever erected by the genius of man. This will be made evident by a single comparison with the Britannia Tubular Bridge erected by Stephenson over the Menai Straits. He adopted the tubular principle, because he believed that the suspension principle could not be made practical for railway traffic, although he had to deal with spans not greater than 470 feet. He built a structure that contained 10,540 tons of iron, and cost 601,000 pounds sterling, or about $3,000,000. Fortunately he has left a calculation on record as to the possible extension of the tubular girder, showing that it would reach the limits in which it could bear only its own weight 68 (62,000 tons), at 1,570 feet. Now, for a span of 1,595½ feet, the Brooklyn Bridge contains but 6,740 tons of material, and will sustain seven times its own weight. Its cost is $9,000,000, whereas a tubular bridge for the same span would contain ten times the weight of metal, and though costing twice as much money, would be without the ability to do any useful work.

Roebling, therefore, solved the problem which had defied Stephenson; and upon his design has been built a successful structure, at half the cost of a tubular bridge that would have fallen when loaded in actual use. It is impossible to furnish any more striking proof of the genius which originated, and of the economy which constructed this triumph of American engineering.

We have thus a monument to the public spirit of the two cities, created by an expenditure as honest and as economical as the management which gave us the Erie Canal, the Croton Aqueduct, and the Central Park. Otherwise, it would have been a monument to the eternal infamy of the trustees and of the engineers 69 under whose supervision it has been erected, and this brings me to the final consideration which I feel constrained to offer on this point.

During all these years of trial, and false report, a great soul lay in the shadow of death, praying only to stay long enough for the completion of the work to which he had devoted his life. I say a great soul, for in the spring-time of youth, with friends and fortune at his command, he gave himself to his country, and for her sake braved death on many a well-fought battle-field. When restored to civil life, his health was sacrificed to the duties which had devolved upon him, as the inheritor of his father's fame, and the executor of his father's plans. Living only for honor, and freed from the temptations of narrow means, how is it conceivable that such a man—whose approval was necessary to every expenditure—should, by conniving with jobbers, throw away more than the life which was dear to him, that he might fulfill his destiny, and leave to his children the heritage of a good name and the glory of a grand achievement? Well may this suffering hero quote the 70 words of Hyperion: "Oh, I have looked with wonder upon those, who, in sorrow and privation, and bodily discomfort, and sickness, which is the shadow of death, have worked right on to the accomplishment of their great purposes; toiling much, enduring much, fulfilling much; and then, with shattered nerves, and sinews all unstrung, have laid themselves down in the grave, and slept the sleep of death, and the world talks of them while they sleep! And as in the sun's eclipse we can behold the great stars shining in the heavens, so in this life-eclipse have these men beheld the lights of the great eternity, burning solemnly and forever!"

And now what is to be the outcome of this great expenditure upon the highway which unites the two cities, for which Dr. Storrs and I have the honor to speak to-day? That Brooklyn will gain in numbers and in wealth with accelerated speed is a foregone conclusion. Whether this gain shall in any wise be at the expense of New York, is a matter in regard to which the great metropolis does not concern herself. Her citizens are content with the knowl 71 edge that she exists and grows with the growth of the whole country, of whose progress and prosperity she is but the exponent and the index. Will the Bridge lead, as has been forcibly suggested, and in some quarters hopefully anticipated, to the further union of the two cities under one name and one government? This suggestion is in part sentimental and in part practical. So far as the union in name is concerned, it is scarcely worth consideration, for in any comparison which our national or local pride may institute between this metropolis and the other great cities of the world, its environment, whether in Long Island, Staten Island, or New Jersey, will always be included. In considering the population of London, no one ever separates the city proper from the surrounding parts. They are properly regarded as one homogeneous aggregation of human beings.

It is only when we come to consider the problem of governing great masses that the serious elements of the question present themselves, and must be determined before a satisfactory answer can be given. The tendency of modern 72 civilization is towards the concentration of population in dense masses. This is due to the higher and more diversified life, which can be secured by association and co-operation on a large scale, affording not merely greater comfort and often luxury, but actually distributing the fruits of labor on a more equitable basis than is possible in sparsely settled regions and among feeble communities. The great improvements of our day in labor-saving machinery, and its application to agriculture, enable the nation to be fed with a less percentage of its total force thus applied, and leave a larger margin of population free to engage in such other pursuits as are best carried on in large cities.

The disclosures of the last census prove the truth of this statement. At the first census in 1790 the population resident in cities was 3.3 per cent. of the total population. This percentage slowly gained at each successive census, until in 1840 it had reached 8.5 per cent. In fifty years it had thus gained a little over five per cent. But in 1850 it rose to 12.5 per cent.; in 1860 it was 16.1 per cent.; in 1870 it 73 was 20.9 per cent., having in this one decade gained as much as in the first fifty years of our political existence. In 1880 the population resident in cities was 22.5 per cent. of the whole population.

With this rapid growth of urban population, have grown the contemporaneous complaints of corrupt administration and bad municipal government. The outcry may be said to be universal, for it comes from both sides of the Atlantic; and the complaints appear to be in direct proportion to the size of cities. It is obvious, therefore, that the knowledge of the art of local government has not kept pace with the growth of population. I am here by your favor to speak for the city of New York, and I should be the last person to throw any discredit on its fair fame; but I think I only give voice to the general feeling, when I say that the citizens of New York are satisfied neither with the structure of its government, nor with its actual administration, even when it is in the hands of intelligent and honest officials. Dissatisfied as we are, no man has been able to devise a system 74 which commends itself to the general approval, and it may be asserted that the remedy is not to be found in devices for any special machinery of government. Experiments without number have been tried, and suggestions in infinite variety have been offered, but to-day no man can say that we have approached any nearer to the idea of good government, which is demanded by the intelligence and the wants of the community.

If, therefore, New York has not yet learned to govern itself, how can it be expected to be better governed by adding half a million to its population, and a great territory to its area, unless it be with the idea that a "little leaven leaveneth the whole lump." Is Brooklyn that leaven? If not, and if possibly "the salt has lost its savor, wherewith shall it be salted?" Brooklyn is now struggling with this problem, it remains to be seen with what success; but meanwhile it is idle to consider the idea of getting rid of our common evils by adding them together.

Besides it is a fundamental axiom in politics, 75 approved by the experience of older countries, as well as of our own, that the sources of power should never be far removed from those who are to feel its exercise. It is the violation of this principle which produces chronic revolution in France, and makes the British rule so obnoxious to the Irish people. This evil is happily avoided when a natural boundary circumscribes administration within narrow limits. While, therefore, we rejoice together at the new bond between New York and Brooklyn, we ought to rejoice the more, that it destroys none of the conditions which permit each city to govern itself, but rather urges them to a generous rivalry in perfecting each its own government, recognizing the truth, that there is no true liberty without law, and that eternal vigilance, which is the only safeguard of liberty, can best be exercised within limited areas.

It would be a most fortunate conclusion, if the completion of this Bridge should arouse public attention to the absolute necessity of good municipal government, and recall the only principle upon which it can ever be successfully 76 founded. There is reason to hope that this result will follow, because the erection of this structure shows how a problem, analogous to that which confronts us in regard to the city government, has been met and solved in the domain of physical science.

The men who controlled this enterprise at the outset were not all of the best type; some of them, as we have seen, were public jobbers. But they knew that they could not build a bridge, although they had no doubt of their ability to govern a city. They thereupon proceeded to organize the knowledge which existed as to the construction of bridges; and they held the organization thus created responsible for results. Now, we know that it is at least as difficult to govern a city as to build a bridge, and yet, as citizens, we have deliberately allowed the ignorance of the community to be organized for its government, and we then complain that it is a failure. Until we imitate the example of the Ring, and organize the intelligence of the community for its government, our complaint is childish and unreasonable. But we 77 shall be told that there is no analogy between building a bridge and governing a city. Let us examine this objection. A city is made up of infinite interests. They vary from hour to hour, and conflict is the law of their being. Many of the elements of social life are what mathematicians term "variables of the independent order." The problem is, to reconcile these conflicting interests and variable elements into one organization which shall work without jar, and allow each citizen to pursue his calling, if it be an honest one, in peace and quiet.

Now, turn to the Bridge. It looks like a motionless mass of masonry and metal; but, as a matter of fact, it is instinct with motion. There is not a particle of matter in it which is at rest even for the minutest portion of time. It is an aggregation of unstable elements, changing with every change in the temperature, and every movement of the heavenly bodies. The problem was, out of these unstable elements, to produce absolute stability; and it was this problem which the engineers, the organized intelligence, had to solve, or confess to inglorious 78 failure. The problem has been solved. In the first construction of suspension bridges it was attempted to check, repress and overcome their motion, and failure resulted. It was then seen that motion is the law of existence for suspension bridges, and provision was made for its free play. Then they became a success. The Bridge before us elongates and contracts between the extremes of temperature from 14 to 16 inches; the vertical rise and fall in the centre of the main span ranges between 2 ft. 3 in. and 2 ft. 9 in.; and before the suspenders were attached to the cable it actually revolved on its own axis through an arc of thirty degrees, when exposed to the sun shining upon it on one side. You do not perceive this motion, and you would know nothing about it unless you watched the gauges which record its movement.

Now if our political system were guided by organized intelligence, it would not seek to repress the free play of human interests and emotions, of human hopes and fears, but would make provision for their development and exercise, in accordance with the higher law of liberty 79 and morality. A large portion of our vices and crimes are created either by law, or its maladministration. These laws exist because organized ignorance, like a highwayman with a club, is permitted to stand in the way of wise legislation and honest administration, and to demand satisfaction from the spoils of office, and the profits of contracts. Of this state of affairs we complain, and on great occasions the community arises in its wrath, and visits summary punishment on the offenders of the hour, and then relapses into chronic grumbling until grievances sufficiently accumulate to stir it again to action.

What is the remedy for this state of affairs? Shall there be no more political parties, and shall we shatter the political machinery which, bad as it is, is far better than no machinery at all? Shall we embrace nihilism as our creed, because we have practical communism forced upon us as the consequence of jobbery, and the imposition of unjust taxes?

No, let us rather learn the lesson of the Bridge. Instead of attempting to restrict suffrage, let us try to educate the voters; instead of disbanding parties, let each citizen within the party always vote, but never for a man who is unfit to hold office. Thus parties, as well as voters, will be organized on the basis of intelligence.

But what man is fit to hold office? Only he who regards political office as a public trust, and not as a private perquisite to be used for the pecuniary advantage of himself or his family, or even his party. Is there intelligence enough in these cities, if thus organized within the parties, to produce the result which we desire? Why, the overthrow of the Tweed Ring was conclusive evidence of the preponderance of public virtue in the city of New York. In no other country in the world, and in no other political system than one which provides for and secures universal suffrage, would such a sudden and peaceful revolution have been possible. The demonstration of this fact was richly worth the twenty-five or thirty millions of dollars which the thieves had stolen. Thereafter, and thenceforth, there could be no doubt whether our city population, heterogeneous as it is, contains within itself sufficient virtue for its own preservation. Let it never be forgotten that the remedy is complete; that it is ever present; that no man ought to be deprived of the opportunity of its exercise; and that, if it be exercised, the will of the community can never be paralyzed. Our safety and our success rest on the ballot in the hands of freemen at the polls, deliberately deposited, never for an unworthy man, but always with a profound sense of the responsibility which should govern every citizen in the exercise of this fundamental right.

If the lesson of the Bridge, which I have thus sought to enforce, shall revive the confidence of the people in their own power, and induce them to use it practically for the election to office of good men, clothed, as were the engineers, with sufficient authority, and held, as they were, to corresponding responsibility for results, then, indeed, will its completion be a public blessing, worthy of the new era of industrial development in which it is our fortunate lot to live.

Great, indeed, has been our national progress. Perhaps we, who belong to a commercial community, do not fully realize its significance and promise. We buy and sell stocks, without stopping to think that they represent the most astonishing achievements of enterprise and skill in the magical extension of our vast railway system; we speculate in wheat, without reflecting on the stupendous fact that the plains of Dakota and California are feeding hungry mouths in Europe; we hear that the Treasury has made a call for bonds, and forget that the rapid extinction of our national debt is a proof of our prosperity and patriotism, as wonderful to the world as was the power we exhibited in the struggle which left that apparently crushing burden upon us. If, then, we deal successfully with the evils which threaten our political life, who can venture to predict the limits of our future wealth and glory—wealth that shall enrich all; glory that shall be no selfish heritage, but the blessing of mankind? Beyond all legends of oriental treasure, beyond all dreams of the golden age, will be the splendor, and majesty, and happiness of the free people dwelling upon this fair domain, when fulfilling the promise of 83 the ages and the hopes of humanity they shall have learned how to make equitable distribution among themselves of the fruits of their common labor. Then, indeed, will be realized by a waiting world the youthful vision of our own Bryant:

"Here the free spirit of mankind at length
Throws its last fetters off; and who shall place
A limit to the giant's untamed strength,
Or curb its swiftness in the forward race?
Far, like the comet's way through infinite space,
Stretches the long untraveled path of light
Into the depths of ages; we may trace
Distant, the brightening glory of its flight,
Till the receding rays are lost to human sight."

At the ocean gateway of such a nation well may stand the stately figure of "Liberty enlightening the World;" and, in hope and faith, as well as gratitude, we write upon the towers of our beautiful Bridge, to be illuminated by her electric ray, the words of exultation, "*Finis coronat opus.*"

84

ORATION

OF

Richard S. Storrs, D.D., LL.D.

Mr. Chairman—Fellow-Citizens: It can surprise no one that we celebrate the completion of this great work, in which lines of delicate and aerial grace are combined with a strength more enduring than of marbles, and the woven wires prolong to these heights the metropolitan avenues. After delays which have often disturbed the popular patience, and have oftener disappointed the hopes of the builders, we gratefully welcome this superb consummation: rejoicing to know that "the silver streak" which so long has divided this city from the continent, is conquered, henceforth, by the silver band stretching above it, careless alike of wind and tide, of ice and fog, of current and of calm. 85

To the mind which, for fourteen years, has watched, guided, and governed the work, looking out upon it through physical organs almost fatally smitten in its prosecution, we bring our eager and unanimous tribute of honor and applause. He who took up, elaborated, and has brought to fulfillment the plans of the father whose own life had been sacrificed in their furtherance, has built to both the noblest memorial. He may with truth have said, heretofore, as the furnaces have glowed from which this welded network has come, in the words of Schiller's "Lay of the Bell:"

"Deep hid within the nether cell
What Force with Fire is moulding thus,
In yonder airy towers shall dwell,
And witness wide and far of us."

He may, at this hour, add for himself the lines which the poet hears from the lips of his House-Master:

"My house is built upon a rock,
And sees unmoved the stormy shock
Of waves that fret below."

It must be a superlative moment in life when one stands on a structure as majestic as this which was at first a mere thought in the brain, which was afterward a plan on the paper, and which has been transported hither, from quarry and mine, from wood-yard and workshop, on the point of his pencil.

He would be the first to acknowledge also, if he were speaking, the intelligent, faithful, indefatigable service rendered in execution of his plans by those who have been associated with him, as assistant engineers, as master mechanics, or as trained, trusted, and experienced workmen. On their knowledge and vigilance, their practiced skill and patient fidelity, the work has of necessity largely depended for its completed grace and strength. They have wrought the zealous labor of years into all parts of it; and it will bear to them hereafter, as it does to-day, most honorable witness.

Some of our honored fellow-citizens, who have borne a distinguished part in this enterprise, are no more here to share our festivities. Mr. John H. Prentice, for years the Treasurer of the Board, wise in counsel, of a liberal yet a watchful economy, of incorruptible integrity, passed from the earth two years ago; but to those who knew him his memory is as fresh as the verdure above his grave at Greenwood. More lately, one who had been from the outset associated with what to many appeared this visionary plan, to whose capacity and experience, his legal skill, his legislative influence, his social distinction, the work has been always largely indebted, and who was for years the President of the Board, has followed into the silent land. It is a grief to all who knew him that he is not here to see the consummation of labors and plans which for years had occupied his life. But his face and figure are before us, almost as distinctly as if he were present; and it will be only the dullest forgetfulness which can ever cease to connect with this Bridge the name of the accomplished scholar, the experienced diplomatist, the untiring worker, the cordial and ever-helpful friend, Mr. Henry C. Murphy.

But others remain to whom the work has brought its burdens, of labor, care, and long solicitude, sometimes, no doubt, of a public criticism whose imperious sharpness they may have felt, but who have followed their plans to completion, without wavering or pause; who have, indeed, expanded those plans as the progress of the work has suggested enlargement; and who, to-day, enter the reward which belongs to those who, after promoting a magnificent enterprise, see it accomplished. Among them are two who were associated with it at the beginning, and who have continued so associated from that day to this—Mr. William C. Kingsley, Mr. James S.T. Stranahan. The judgment cannot be mistaken which affirms that to these men, more than to any other citizens remaining among us, the prosecution of this work to its crowning success is properly ascribed. They are the true orators of the hour. We may praise, but they have builded. On the tenacity of their purpose, of which that of these combining wires only presents the physical image,—on the lift of their wills, stronger than of these consenting cables,—the immense structure has risen to its place. No grander work has it been given to men to do for the city, which will feel the unfailing impulse of their foresight and courage, their 89 wisdom in counsel, and their resolute service, to the end of its history!

Mr. William Marshall, Gen. Henry W. Slocum, were also connected with the work at the outset, and, with intervals in the period of their service, have given it important assistance to the end; while others are with us who have joined with intelligence, enthusiasm, and helpfulness in the councils of the Board at different times. We rejoice in the presence of all those who, earlier or later, have taken part in the plans, at once vast and minute, which now are realized. We offer them the tribute of our admiring and grateful esteem. We trust that their remembrance of the work they have accomplished, and their personal experience of its manifold benefits, may continue through many happy years. And we congratulate ourselves, as well as them, that the city will keep the memorial of them, not in yonder tablets alone, but in the great fabric above which those stand, while stone and steel retain their strength.

But, after all, the real builder of this surpassing and significant structure has been the people: 90 whose watchfulness of its progress has been constant, whose desire for its benefits has been the incentive behind its plans, by whom its treasury has been supplied, whose exultant gladness now welcomes its success. The people of New York have illustrated anew their magnanimous spirit in cheerfully supplying their share of the cost, though not anticipating from such large outlay direct reliefs and signal advantages. The people of

Brooklyn have shown at least an intelligent, intrepid, and farsighted sagacity, in readily accepting the immediate burdens in expectation of future returns.

Such a popular achievement is one to be proud of. St. Petersburg could be commenced 180 years ago—almost to a day, on May 27th, 1703—and could afterward be built, by the will of an autocrat, to give him a new centre of empire, with a nearer outlook over Europe; its palaces rising on artificial foundations, which it cost, it is said, 100,000 lives in the first year to lay. Paris could be reconstructed, twenty-five years ago, by the mandate of an emperor, determined to make it more beautiful than before, to 91 open new avenues for guns and troops, to give to its laborers, who might become troublesome, desired occupation. But not only have these cities of ours been founded, built, reconstructed by the people, but this charming and mighty avenue in the air, by which they are henceforth rebuilt into one, is to the people's honor and praise. It shows what multitudes, democratically organized, can do if they will. It will show, to those who shall succeed us, to what largeness of enterprise, what patience of purpose, what liberal wisdom, the populations now ruling these associated cities were competent in their time. It takes the aspect, as so regarded, of a durable monument to Democracy itself.

We congratulate the Mayors of both the cities, with their associates in the government of them, on the public spirit manifested by both, on the ampler opportunities offered to each, and on those intimate alliances between them which are a source of happiness to both, and which are almost certainly prophetic of an organic union to be realized hereafter. And we trust that the crosses, encircled by the laurel wreath, on the 92 original seal of New Amsterdam, with the Dutch legend of this city, "Union makes Strength," may continue to describe them, whether or not stamped upon parchments and blazoned on banners, as long as human eyes shall see them.

The work now completed is of interest to both cities, and its enduring and multiplying benefits will be found, we are confident, to be common, not local.

We who have made and steadfastly kept our homes in Brooklyn, and who are fond and proud of the city—for its fresh, bracing, and healthful air, and the brilliant outstretch of sea and land which opens from its Heights; for its scores of thousands of prosperous homes; for its unsurpassed schools, its co-operating churches, the social temper which pervades it, the independence and enterprise of its journals, and the local enthusiasms which they fruitfully foster; for its general liberality, and the occasional splendid examples of individual munificence which have given it fame; for its recent but energetic institutions, of literature, art, and a noble philanthropy; and for the stimulating enterprise and 93 culture of the young life which is coming to command in it—we have obvious reason to rejoice in the work which brings us into nearer connection with all that is delightful and all that is enriching in the metropolis, and with that diverging system of railways, overspreading the continent, which has in the commercial capital its natural centre of radiation.

We have no word of criticism to speak, only words of most hearty admiration, for the safe and speedy water-service on the lines of the ferries which has given us heretofore such easy transportation from city to city, without delays that were not unavoidable, and with remarkable exemption from disaster. So far as human carefulness and skill could assure safety and speed, in the midst of conditions unfriendly to both, the management of these ferries has been peerless, their success unsurpassed. To them is due, in largest measure, the rapid growth already here realized. They have formed the indispensable arteries, of supply and transmission, through which the circulating life-blood has flowed, and their ministry to this city has been constant and 94 vital. But we confess ourselves glad to reach, with surer certainty and a greater rapidity, the libraries and galleries, the churches and the homes, as well as the resorts of business and of pleasure, with which we are now in instant connection; and the horizon widens around us as we touch with more immediate contact the lines of travel which open hence to the edges of the continent.

If we have not as much to offer in immediate return, we have, at least, a broad expanse of uncovered acres within the city, for the easy occupation of those who wish homes, either modest or splendid, or who shall wish such as the growth of the metropolis multiplies its population into the millions, crowds its roofs higher toward the stars, and makes a productive silver mine of each several house-lot. And to those who visit us but at intervals we can open not only yonder park, set like an emerald in the great circular sweep of our boundaries from the waters of the Narrows to the waters of the Sound, but also their readiest approach to the ocean. The capital and the sea are henceforth 95 brought to nearer neighborhood. Long Island bays, and brooks, and beaches, are within readier reach of the town. The winds that have touched no other land this side of Cuba are more accessible to those who seek their tonic breath. The long roll of the surf on the shore breaks closer than before to office and mansion, and to tenement chamber.

The benefits will, therefore, be reciprocal, which pass back and forth across this solid and stately frame-work; and both cities will rejoice, we gladly hope, in the patience and labor, the disciplined skill, the large expenditure, of which it is the trophy and fruit. New York has now the unique opportunity to widen its boundaries to the sea, and around its brilliant civic shield, more stately and manifold than that of Achilles, by the aid of those who have wrought already these twisted bracelets and clasping cables, to set the glowing margin of the Ocean-stream.

This work is important, too, we cannot but feel, in wider relations; for what it signifies, as for what it secures, and for all that it promises. Itself a representative product and part of the 96 new civilization, one standing on it finds an outlook from it of larger circumference than that of these cities.

Every enterprise like this, successfully accomplished, becomes an incentive to others like it. It leads on to such, and supplies incessant encouragement to them. We may not know, or probably conjecture, what these are to be, in the city or the State, in the years that shall come. But, whatever they may be, for the more complete equipment of either with conditions of happiness and the instruments of progress, they will all take an impulse from that which here has been accomplished. Such a trophy of triumph over an original obstacle of Nature will not contribute to sleep in others; and whatever is needed of material improvement, throughout the State of which it is our pride to be citizens, will be only more surely and speedily supplied because of this impressive success.

It is, therefore, most fitting to our festival that we are permitted to welcome to it the Chief Magistrate of the State, with those representing its different regions in the legislative councils. 97 We rejoice to remember that the work before us has been assisted by the favoring action of those heretofore in authority in the State; and we trust that to those now holding high offices in it, who are present to-day, the occasion will be one of pleasant experience, and of enlarged and reinforced expectation.

Indeed, it is not extravagant to say that the future of the country opens before us, as we see what skill and will can do to overleap obstacles, and make nature subservient to human designs. So we gladly welcome these eminent men from other States; while the presence of the Executive Head of the Nation, and of some of the members of his Cabinet, is appropriate to the time, as it is an occasion of sincere and profound gratification to us all. Without the concurrence of the National Government, this structure, though primarily of local relations, as reaching across these navigable waters, could not have been built. We feel assured that those honorably representing that Government, who favor its completion with their attendance, and in whose presence political differences are 98 forgotten, will share with us in the joyful pride with which we regard it, and in the inspiring anticipation that the physical apparatus of civilization in the land is to take fresh impulse, not impediment or hindrance, from that which here has been effected. The day seems brought distinctly nearer when the Nation, equipped with the latest implements furnished by science, shall master and use as never before its rich domain.

Not only the modern spirit is here, even in eminence, which dares great effort for great advantage; but the chiefest of modern instruments is here, which is the ancient untractable iron, transfigured into steel.

It was a sign, and even a measure, of ancient degeneracy, when the age of Gold was followed if not forgotten by one of Iron. Decadence of arts, of learning and laws, of society itself, was implied in the fact. The more intrepid intelligence, the more versatile energy, amid which we live, have achieved the success of combining the two: so that while it is true now, as of old, that "no mattock plunges a golden edge into the ground, and no nail drives a silver point into the plank," it is also true that, under the stimulus of the larger expenditure which the added supplies of gold make possible, the duller metal has taken a fineness, a brightness and hardness, with a tensile strength, before unfamiliar.

The iron, as of old, quarries the gold, and cuts it out from river-bed and from rock. But, under the alchemy which gold applies, the iron takes nobler properties upon it. Converted into steel, in masses that would lately have staggered men's thoughts, it becomes the kingliest instrument of peoples for subduing the earth. Things dainty and things mighty are fashioned from it in equal abundance:—gun-carriage and cannon, with the solid platforms on which they rest; the largest castings, and heaviest plates, as well as wheel, axle, and rail, as well as screw or file or saw. It is shaped into the hulls of ships. It is built alike into column and truss, balcony, roof, and springing dome. To the loom and the press, and the boiler from whose fierce and untiring heart their force is supplied, it is equally apt; while, as drawn into delicate wires, it is coiled into springs, woven into gauze, sharpened into needles, twisted into ropes; it is made to yield music in all our homes; electric currents are sent upon it, along our streets, around the world; it enables us to talk with correspondents afar, or it is knit, as before our eyes, into the new and noble causeways of pleasure and of commerce.

I hardly think that we yet appreciate the significance of this change which has passed upon iron. It is the industrial victory of the century, not to have heaped the extracted gold in higher piles, or to have crowded the bursting vaults with accumulated silver, but to have conferred, by the sovereign touch of scientific invention, flexibility, grace, variety of use, an almost ethereal and spiritual virtue, on the stubbornest of common metals. The indications of physical achievement in the future, thus inaugurated, outrun the compass of human thought.

Two bridges lie near each other, across the historical stream of the Moldau, under the shadow of the ancient and haughty palace at Prague—the one the picturesque bridge of St. Nepomuk, patron of bridges throughout Bohemia, of massive stone, which occupied a century and a half in its erection, and was finished almost four centuries ago, with stately statues along its sides, with a superb monument at its end, sustaining symbolic and portrait figures; the other an iron suspension-bridge, built and finished in three years, a half century since, and singularly contrasting, in its lightness and grace, the sombre solidity of the first. It is impossible to look upon the two without feeling how distinctly the different ages to which they belong are indicated by them, and how the ceremonial and military character of the centuries that are past has been superseded by the rapid and practical spirit of commerce.

But the modern bridge is there a small one, and rests at the centre on an island and a pier. The structure before us, the largest of its class as yet in the world, in its swifter, more graceful, and more daring leap from bank to bank, across the tides of this arm of the sea, not only illustrates the bolder temper which is natural here, the readiness to attempt unparalleled works, 102 the disdain of difficulties in unfaltering reliance on exact calculation, but, in the material out of which it is wrought, it shows the new supremacy of man over the metal which, in former time, he scarcely could use save for rude and coarse implements. The steel of the blades of Damascus or Toledo is not here needed; nor that of the chisel, the knife-blade, the watch-spring, or the surgical instrument. But the steel of the mediæval lance-head or sabre was hardly finer than that which is here built into a Castle, which the sea cannot shake, whose binding cement the rains cannot loosen, and before whose undecaying parapets open fairer visions of island and town, of earth, water, and sky, than from any fortress along the Rhine. There is inexhaustible promise in the fact.

Of course, too, there is impressively before us—installed as on this fair and brilliant civic throne—that desire for swiftest intercommunication between towns and districts divided from each other, which belongs to our times, and which is to be an energetic, enduring, and salutary force in moulding the nation. 103

The years are not distant in which separated communities regarded each other with aversion and distrust, and the effort was mutual to raise barriers between them, not to unite them in closer alliance. Now, the traffic of one is vitally dependent on the industries of the other; the counting-room in the one has the factory or the warehouse tributary to it established in the other; and the demand is imperative that the two be linked, by all possible mechanisms, in a union as complete as if no chasm had opened between them. So these cities are henceforth united; and so all cities, which may minister to each other, are bound more and more in intimate combinations. Santa Fé, which soon celebrates the third of a millenium since its foundation, reaches out its connections toward the newest log-city in Washington Territory; and the oldest towns upon our seaboard find allies in those that have risen, like exhalations, along the Western lakes and rivers.

This mighty and symmetrical band before us seems to stand as the type of all that immeasurable communicating system which is more 104 completely with every year to interlink cities, to confederate States, to make one country of our distributed imperial domain, and to weave its history into a vast, harmonious contexture, as messages fly instantaneously across it, and the rapid trains rush back and forth, like shuttles upon a mighty loom.

It is not fanciful, either, to feel that in all its history, and in what is peculiar in its constitution, it becomes a noble, visible symbol of that benign Peace amid which its towers and roadway have risen, and which, we trust, it may long continue to signalize and to share.

We may look at this moment on the site of the ship-yard from which, in March, 1862, twenty-one years ago, went forth the unmasted and raft-like "Monitor," with its flat decks, its low bulwarks, its guarded mechanism, its heavy armament, and its impenetrable revolving turret, to that near battle with the "Merrimac," on which, as it seemed to us at the time, the destiny of the nation was perilously poised. The material of which the ship was wrought was largely that which is built in beauty into this luxurious lofty 105 fabric. But no contrast could be greater among the works of human genius than between the compact and rigid solidity into which the iron had there been forged and wedged and rammed, and these waving and graceful curves, swinging downward and up, almost like blossoming festooned vines along the perfumed Italian lanes; this alluring roadway, resting on towers which rise like those of ancient cathedrals; this lace-work of threads, interweaving their separate delicate strengths into the complex solidity of the whole.

The ship was for war, and the Bridge is for peace: — the product of it; almost, one might say, its express palpable emblem, in its harmony of proportions, its dainty elegance, its advantages for all, and its ample convenience. The deadly raft, floating level with waves, was related to this ethereal structure, whose finest curves are wrought in the strength of toughest steel. We could not have had this except for that unsightly craft, which at first refused to be steered, which bumped headlong against our piers, which almost sank while being towed to the field of its fame, 106 and which, at last, when its mission was fulfilled, found its grave in the deep over whose waters, and near their line, its shattering lightnings had been shot. This structure will stand, we fondly trust, for generations to come, even for centuries, while metal and granite retain their coherence; not only emitting, when the wind surges or plays through its network, that aerial music of which it is the mighty harp, but representing to every eye the manifold bonds of interest and affection, of sympathy and purpose, of common political faith and hope, over and from whose mightier chords shall rise the living and unmatched harmonies of continental gladness and praise.

While no man, therefore, can measure in thought the vast processions — 40,000,000 a year, it already is computed — which shall pass back and forth across this pathway, or shall pause on its summit to survey the vast and bright panorama, to greet the break of summer-morning, or watch the pageant of closing day, we may hope that the one use to which it never will need to be put is that of war; that the one tramp not 107 to be heard on it is that of soldiers marching to battle; that the only wheels whose roll it shall not be called to echo are the wheels of the tumbrils of troops and artillery. Born of peace, and signifying peace, may its mission of peace be uninterrupted, till its strong towers and cables fall!

If such expectations shall be fulfilled, of mechanical invention ever advancing, of cities and States linked more closely, of beneficent peace assured to all, it is impossible to assign any limit to the coming expansion and opulence of these cities, or to the influence which they shall exert on the developing life of the country.

Cities have often, in other times, been created by war; as men were crowded together in them the better to escape the whirls of strife by which the unwalled districts were ravaged, or the more effectively to combine their force against threatening foes. And it is a striking suggestion of history that to the frightful ravages of the Huns—swarthy, ill-shaped, ferocious, destroying—may have been due the Great Wall of China, for the protection of its remote towns, as to them, on the other hand, was certainly due the foundation of Venice. The first inhabitants of what has been since that queenly city—along whose liquid and level streets the traveler passes, between palaces, churches, and fascinating squares, in constant delight—its first inhabitants fled before Attila, to the flooded lagoons which were afterward to blossom into the beauty of a consummate art. The fearful crash of blood and fire in which Aquileia and Padua fell smote Venice into existence.

But even the city thus born of war must afterward be built up by peace, when the strifes which had pushed it to its sudden beginning had died into the distant silence. The fishing industry, the manufacture of salt, the timid commerce, gradually expanding till it left the rivers and sought the sea, these, with other related industries, had made Venetian galleys known on the eastern Mediterranean before the immense rush of the crusades crowded tumultuously over its quays and many bridges. Its variety of industry, and its commercial connections, turned that vast movement into another source of wealth. It rose rapidly to that naval supremacy which enabled it to capture piratical vessels and wealthy galleons, to seize or sack Ionian cities, to storm Byzantium, and make the south of Greece its suburb. Its manufactures were multiplied. Its dockyards were thronged with busy workmen. Its palaces were crowded with precious and famous works of art, while themselves marvels of beauty. St. Mark's unfolded its magnificent loveliness above the great square. In the palace adjoining was the seat of a dominion at the time unsurpassed, and still brilliant in history; and it was in no fanciful or exaggerated pride that the Doge was wont yearly, on Ascension Day, to wed the Adriatic with a ring, as the bridegroom weds the bride.

Dreamlike as it seems, equally with Amsterdam, the larger and richer "Venice of the North," it was erected by hardy hands. The various works and arts of peace, with a prosperous commerce, were the real piles, sunken beneath the flashing surface, on which church and palace, piazza and arsenal, all arose. It was only when these unseen supports secretly failed 110 that advancement ceased, and the horses of St. Mark at last were bridled. Not all the wars, with Genoa, Hungary, with Western Europe, the Greek Empire, or the Ottoman—not earthquake, plague, or conflagration, though by all it was smitten—overwhelmed the city whose place in Europe had been so distinguished. The decadence of enterprise, the growing discredit put upon industry, the final discovery by Vasco da Gama of the passage around the Cape of Good Hope, diverting traffic into new channels—these laid their silent and tightening grasp on the power of Venice, till

"the salt sea-weed
Clung to the marble of her palaces,"

and the glory of the past was merged in a gloom which later centuries have not lightened. There is a lesson and a promise in the fact.

New York itself may almost be said to have sprung from war; as the vast excitements of the forty years' wrestle between Spain and its revolted provinces gave incentive, at least, to the settlement of New Netherland. But the city, since its real development was begun, has been 111 almost wholly built up by peace; and the swiftness of its progress in our own time, which challenges parallel, shows what, if the ministry of this peace shall continue, may be looked for in the future.

When the Dutch traders raised their storehouse of logs on yonder untamed and desolate strand, perhaps as early as 1615; when the Walloons established their settlement on this side of the river, in 1624, at that "Walloons' Bay" which we still call the Wallabout; or when, later, in 1626, Manhattan Island, estimated to contain 22,000 acres, was purchased from the Indians for $24, paid in beads, buttons and trinkets, and the Block House was built, with cedar palisades, on the site of the Battery, it is, of course, commonplace to say that they who had come hither could scarcely have had the least conception of what a career they thus were commencing for two great cities. But it is not so wholly commonplace to say that those who saw this now wealthy and splendid New York a hundred years since, less conspicuous than Boston, far smaller than Philadelphia, with its first bank established 112 in 1784, and not fully chartered till seven years later; with its first daily paper in 1785; its first ship in the Eastern trade returning in May of the same year; its first Directory published in 1786, and containing only 900 names; its Broadway extending only to St. Paul's; with the grounds about Reade street grazing-fields for cattle, and with ducks still shot in that Beekman's Swamp which the traffic in leather has since made famous: or those who saw it even fifty years ago, when its population was little more than one-third of the present population of this younger city; when its first Mayor had not been chosen by popular election; when gas had but lately been introduced, and the superseding of the primitive pumps by Croton water had not yet been projected—they, all, could hardly have imagined what already the city should have become: the recognized centre of the commerce of the Continent; one of the principal cities of the world.

So those who have lived in this city from childhood, and who hardly yet claim the dignities of age, could scarcely have conjectured, 113 when looking on what Mr. Murphy recalled as the village of his youth, "a hamlet of a hundred houses," that it should have become, in our time, a city of nearly 70,000 dwelling houses, occupied by twice as many families; with a population, by the census rates, of little less than 700,000; with more than 150,000 children in its public and private schools; with 330 miles of paved streets, as many as last year in New York, and with more than 200 additional miles impatiently waiting to be paved; with 130 miles of street railway track, over which last year 88,000,000 of passengers were carried; with nearly 2,500 miles of telegraph and telephone wire knitting it together; with 35,000,000 of gallons of water, the best on the continent, to which 20,000,000 more are soon to be added, daily distributed in its houses, through 360 miles of pipe; with an aggregate value of real property exceeding certainly $400,000,000; with an annual tax levy of $6,500,000; with manufactures in it whose reported product in 1880 was $103,000,000; with a water-front, of pier, dock, basin, canal, already exceeding 25 miles, and not as yet half developed, 114 at which lies shipping from all the world, more largely than at the piers of New York; and, finally, with what to most modern communities appears to flash as a costly but brilliant diamond necklace, a public debt, beginning now to diminish, it is true, but still approaching, in net amount, $37,500,000!

The child watches, in happy wonder, the swelling film of soapy water into whose iridescent globe he has blown the speck from the bowl of the pipe. But this amazing development around us is not of airy and vanishing films. It is solidly constructed, in marble and brick, in stone and iron, while the proportions to which it has swelled surpass precedent, and rebuke the timidity of the boldest prediction. But that which has built it has been simply the industry, manifold, constant, going on in these cities, to which peace offers incentive and room.

Their future advancement is to come in like manner: not through a prestige derived from their history; not by the gradual increments of their wealth, already collected; not by the riches which they invite to themselves from other cities 115 and distant coasts; not even from their beautiful fortune of location; but by prosperous manufactures prosecuted in them; by the traffic which radiates over the country; by the foreign commerce which, in values increasing every year, seeks this harbor. Each railway whose rapid wheels roll hither, from East or West, from North or South, from the rocks of Newfoundland or the copper-deposits of Lake Superior, from the orange groves of Florida, the Louisiana bayous, the silver ridges of the West, the Golden Gate, gives its guaranty of growth to the still young metropolis. On the cotton fields of the South, and its sugar plantations; on coal mines, and iron mines; on the lakes which winter roofs with ice, and from which drips refreshing coolness through our summer; on fisheries, factories, wheat fields, pine forests; on meadows wealthy with grains or grass, and orchards bending beneath their burdens, this enlarging prosperity must be maintained; and on the steamships, and the telegraph lines, which interweave us with all the world. The swart miner must do his part for it; the ingenious workman, in 116 whatever department; the ploughman in the field, and the fisherman on the banks; the man of science, putting Nature to the question; the laborer, with no other capital than his muscle; the sailor on the sea, wherever commerce opens its wings.

Our Arch of Triumph is, therefore, fitly this Bridge of Peace. Our Brandenburg Gate, bearing on its summit no car of military victory, is this great work of industrial skill. It stands, not, like the Arch famous at Milan, outside the city, but in the midst of these united and busy populations. And if the tranquil public order which it celebrates and prefigures shall continue as years proceed, not London itself, a century hence, will surpass the compass of this united city by the sea, in which all civilized nations of mankind have already their many representatives, and to which the world shall pay an increasing annual tribute.

And so the last suggestion comes, which the hour presents, and of which the time allows the expression.

It was not to an American mind alone that 117 we owed the "Monitor," of which I have spoken, but also to one trained in Swedish schools, the Swedish army, and representing that brave nationality. It is not to a native American mind that the scheme of construction carried out in this Bridge is to be ascribed, but to one representing the German peoples, who, in such enriching and fruitful multitudes, have found here their home. American enterprise, American money, built them both. But the skill which devised, and much, no doubt, of the labor which wrought them, came from afar.

Local and particular as is the work, therefore, it represents that fellowship of the Nations which is more and more prominently a fact of our times, and which gives to these cities incessant augmentation. When, by and by, on yonder island the majestic French statue of Liberty shall stand, holding in its hand the radiant crown of electric flames, and answering by them to those as brilliant along this causeway, our beautiful bay will have taken what specially illuminates and adorns it from Central and from Western Europe. The distant lands from which oceans 118 divide us, though we touch them each moment with the fingers of the telegraph, will have set this conspicuous double crown on the head of our harbor. The alliances of nations, the peace of the world, will seem to find illustrious prediction in such superb and novel regalia.

Friends, and Fellow-Citizens: Let us not forget that, in the growth of these cities, henceforth united, and destined ere long to be formally one, lies either a threat, or one of the conspicuous promises of the time.

Cities have always been powers in history. Athens educated Greece, as well as adorned it, while Corinth filled the throbbing and thirsty Hellenic veins with poisoned blood. The weight of Constantinople broke the Roman Empire asunder. The capture of the same magnificent city gave to the Turks their establishment in Europe for the following centuries. Even where they have not had such a commanding pre-eminence of location, the social, political, moral force proceeding from cities has been vigorous in impression, immense in extent. The passion of Paris, for a hundred years, has created or directed the sentiment of France. Berlin is more than the legislative or administrative centre of the German Empire. Even a government as autocratic as that of the Czar, in a country as undeveloped as Russia, has to consult the popular feeling of St. Petersburg or of Moscow.

In our nation, political power is widely distributed, and the largest or wealthiest commercial centre can have but its share. Great as is the weight of the aggregate vote in these henceforth compacted cities, the vote of the State will always overbear it. Amid the suffrages of the nation at large, it can only be reckoned as one of many consenting or conflicting factors. But the influence which constantly proceeds from these cities—on their journalism, not only, or on the issues of their book-presses, or on the multitudes going forth from them, but on the example presented by them of intellectual, social, religious life—this, for shadow and check, or for fine inspiration, is already of unlimited extent, of incalculable force. It must increase as they expand, and are lifted before the country to a new elevation.

A larger and a smaller sun are sometimes associated, astronomers tell us, to form a binary centre in the heavens, for what is, doubtless, an unseen system receiving from them impulse and light. On a scale not utterly insignificant, a parallel may be hereafter suggested in the relation of these combined cities to a part, at least, of our national system. Their attitude and action during the war—successfully closed under the gallant military leadership of men whom we gladly welcome and honor—were of vast advantage to the national cause. The moral, political, intellectual temper, which dominates in them, as years go on, will touch with beauty, or scar with scorching and baleful heats, extended regions. Their religious life, as it glows in intensity, or with a faint and failing lustre, will be repeated in answering image from the widening frontier. The beneficence which gives them grace and consecration, and which, as lately, they follow to the grave with universal benediction, or, on the other hand, the selfish ambitions which crowd and crush along their streets, intent only on accumulated wealth and its sumptuous display, 121 or the glittering vices which they accept and set on high — these will make their impression on those who never cross the continent to our homes, to whom our journals are but names.

Surely, we should not go from this hour, which marks a new era in the history of these cities, and which points to their future indefinite expansion, without the purpose in each of us, that, so far forth as in us lies, with their increase in numbers, wealth, equipment, shall also proceed, with equal step, their progress in whatever is noblest and best in private and in public life; that all which sets humanity forward shall come in them to ampler endowment, more renowned exhibition: so that, linked together, as hereafter they must be, and seeing "the purple deepening in their robes of power," they may be always increasingly conscious of fulfilled obligation to the Nation and to God; may make the land, at whose magnificent gateway they stand, their constant debtor; and may contribute their mighty part toward that ultimate perfect Human Society for which the seer could find no image so meet or so majestic as that of a City, coming down 122 from above, its stones laid with fair colors, its foundations with sapphires, its windows of agates, its gates of carbuncles, and all its borders of pleasant stones, with the sovereign promise resplendent above it, —

"And great shall be the Peace of thy children!"

www.ingramcontent.com/pod-product-compliance
Lightning Source LLC
Chambersburg PA
CBHW030509220526
45464CB00006B/2722